COSMOLOGY AND BUDDHIST THOUGHT: A CONVERSATION WITH DR. NEIL DEGRASSE TYSON

by

Dr. Jerome Freedman

Dr. Jerome Freedman

Copyright © 2011-2016 Dr. Jerome Freedman

All rights reserved.

ISBN: 1492766763
ISBN-13: 978-1492766766

DEDICATION

This book is dedicated to:

My teacher, Zen Master Thich Nhat Hanh and all the Buddhas and Bodhisattvas before and after him

and

to Dr. Neil deGrasse Tyson, without whose help this would not have been possible.

CONTENTS

DEDICATION III
CONTENTS I
PREFACE III
1 INTRODUCTION 5
2 A SHORT HISTORY OF THE COSMOS 11
    THE BIG BANG 11
    THE SOLAR SYSTEM 12
    THE EARTH 12
    LIFE 13
3 FOUNDATIONS OF BUDDHIST THOUGHT 15
    INTRODUCTION 15
    THE FOUR NOBLE TRUTHS 16
    THE EIGHTFOLD PATH 17
        Right View 18
        Right Speech 19
        Right Thought 19
        Right Action 19
        Right Livelihood 20
        Right Diligence 20
        Right Mindfulness 21
        Right Concentration 21
    CO-DEPENDENT ORIGINATION 22
    INTERBEING 23
    EMPTINESS AND NON-SELF 23
4 THE STORIES WE TELL 25
5 ORIGINS AND THE GOLDILOCKS ZONE 29

6 EXOPLANETS 33
7 CO-DEPENDENT ARISING, INTERBEING, AND
IMPERMANENCE 37
8 TIME SCALES 43
9 BUDDHIST THOUGHT AND THE SELF 48
10 NIRVANA 52
11 ECOLOGY 55
12 CONSCIOUSNESS 57
13 RELIGION 61
14 MINDLESS CONSUMPTION 67
15 DON'T DISPAIR – TAKE ACTION 75
16 CLOSING REMARKS 77
ABOUT THE AUTHOR 79
ACKNOWLEDGMENTS 81

PREFACE

Cosmology and Buddhist Thought is the result of a conversation with astrophysicist and television celebrity, Dr. **Neil deGrasse Tyson**, that took place in New York at the end of May, 2011.

While cosmologists look at the outer space with massive instruments and difficult calculations using mathematics, Buddhists look at inner space with mindfulness and concentration.

Both paths lead to insights about fundamental questions about why and how we got here, what we are doing here, what are our connections and interactions with the universe, and what is our destiny.

Cosmologists study life on earth from an objective point of view and examine its causes. Buddhists study life on earth with regard to understanding its challenges and how to resolve them.

Both Buddhism and cosmology recognize that we live in an emerging, evolving, and impermanent universe – everything is changing.

Dr. Tyson's insights into the origins of life on earth, time scales, ecology, and religion are inspiring, to say the least, and extremely wise to read and contemplate. His responsiveness to Buddhist ideas of co-dependent arising, interbeing (interconnectedness), impermanence, the self, nirvana, and consciousness demonstrates much more than a "Reader's Digest knowledge of Buddhism."

A close reading of the conversation reveals that Buddhist thought does not have much to contribute to cosmology. Cosmology, on the other hand, has a lot to contribute to Buddhism.

1 INTRODUCTION

On May 26, 2011, I was in New York visiting my son. On that day, I had the distinct pleasure of interviewing Dr. Neil deGrasse Tyson.

Dr. Tyson is the Frederick P. Rose Director of the Hayden Planetarium at the Rose Center for Earth and Space as well as a research associate in the department of astrophysics at the American Museum of Natural History in New York City. He is also quite a television celebrity because of his appearances in Origins, The Universe, Through the Wormhole, NOVA, Cosmos: A Personal Journey (Carl Sagan), and in other scientific venues.

My first experience with Dr. Tyson was in 2007 when I happened to catch a glimpse of the TV show, *The Universe* on The History Channel. I was impressed with his vast knowledge and his ability to explain the mysteries of the universe in clear, everyday language.

My interest in cosmology and astronomy was rekindled in 2003 when I first watched Brian Green's *The Elegant Universe* on PBS. I studied physics in graduate school and received an A. M. in physics from Washington University in St. Louis and an M. S. in physics from the University of Chicago.

My primary interest in studying physics was to try to understand God through understanding the nature of the universe. I loved quantum theory and the paradoxical nature of its propositions.

Since I was Jewish on my parents' side, I had tried to find some answers in the Jewish tradition.

All of this was the result of three episodes of transcendental experiences I had as an undergraduate student. Every one of these experiences raised more questions about my religious convictions than they answered. In each case, I felt like I lost my personal identity and merged with the oneness of the universe. There was an immense feeling of bliss, happiness, and well-being from each one. These experiences were so powerful that they shaped and inspired almost everything in my life.

The first of these experiences came during the Yom Kipper services in 1959. I thought, for sure, this was an act of God! I really prayed hard for the rest of the day.

I became utterly confused when the second experience occurred during the singing of *Ode to Joy* in a performance of Beethoven's Ninth Symphony by the St. Louis Symphony Orchestra. The theme of *Ode to Joy* is the brotherhood and unity of mankind. The performance was truly magnificent and I was in rapture for an extended period of time.

In a sense, this experience transcended the first because it went beyond being a "Jewish" experience to all humanity, and, as I would now say, to all sentient beings. My Jewish mind became unsettled and disturbed.

Then it happened again. This time it was in the kitchen of our old house on Cornell Avenue. I was sitting around the table with my parents and all my five siblings. Judy, my dearly departed sister, was home from the University of Oklahoma for the weekend. It was one of the few times I felt unconditional love from my family. It didn't matter that no one was paying attention to me at that moment. I was just happy to be there. It probably was the last time we were all be together around that particular table. How were we to know that Judy would soon get married, have two wonderful boys, and die from Leukemia nine years later?

So, we have here, first a "Jewish" experience, then a "universal brotherhood" experience, and finally, a "family" experience. Putting it all together had become the *koan* of my life.

During my first year at the University of Chicago, I met professor Eugene Gendlin, author of *Focusing*, and an absolutely lovely individual. We had several interesting conversations about physics and existential philosophy.

In one of them, I had an insight that could have turned

in determining my life's work, had I been ready for it. The insight had to do with altered states of consciousness similar to the ones described above. To understand it, you have to understand how a laser works.

At that time, lasers were fairly new, and their principles of operation were interesting to me. Laser is an acronym for "light amplification by stimulated emission of radiation." Unlike ordinary light, the light from a laser is coherent. This means that the light emitted from a laser can be sharply focused and very intense.

My insight, which is being researched today, was that altered states of consciousness could possibly be a coherent firing of neurons in the brain in such a way as to stimulate a feeling of euphoria, rapture, bliss, and unity with the whole universe. I felt totally incapable of delving into neuroscience and the physics of the brain on my own, as this would definitely be considered a wacky idea in those days.

During one of my meetings with Dr. Gendlin, I presented a paper I had written for him and his colleagues on *"The Strange Reality of the Quanta."* The paper was broadly accepted and many questions were discussed. I wish I could find my copy of it today.

After completing my masters at the University of Chicago, my attention turned to Hinduism, Buddhism, and other modalities of conscious conduct instead of physics. Physics would often come into my conversations with fellow searchers. For example, when I met Fritjof Capra, author of *The Tao of Physics: An Exploration of the Parallels Between Modern Physics and Eastern Mysticism*, we would have long conversations about the parallels he wrote about.

For many years, I was concerned with raising a family, earning a living, and spiritual practice, not necessarily in that order. With Brian Green's TV program, my interest in physics, astronomy and cosmology merged with my Buddhist studies and practices in a very meaningful way.

Around 2005, I came across the Dalai Lama's book, The *Universe in a Single Atom: The Convergence of Science and*

Spirituality and listened with great interest as Richard Gere read its pages. This, together with the Dalai Lama's meetings with scientists, doctors, psychologists, and philosophers in Mind and Life conferences led to a resurgence of my interest in physics.

Recently, I have engaged in some educational courses in astronomy and cosmology that brought me up to speed with such cosmological concepts as inflation, dark matter, dark energy, the expanding universe, and the life cycle of stars. This has been really wonderful and I've enjoyed it very much.

With my good friend, I have also been attending the Dean Lectures on astronomy at the Morrison Planetarium in the California Academy of Sciences in Golden Gate Park in San Francisco. We have really enjoyed them as well.

By the time I met with Dr. Neil deGrasse Tyson, I felt that I had prepared myself well enough to understand any science he may invoke during our conversation.

Earlier in May of 2011, I had completed my job as an expert witness and software development engineer for the Technical Committee of the Department of Justice monitoring Microsoft's compliance with the anti-trust settlement agreement. The contract began in 2004 and was supposed to only last six weeks. It only ended because the DOJ and the plaintiffs had thought that we had done our job.

I was feeling a wonderful sense of freedom when the contract ended because I had made valuable contributions to the case and earned a good deal of money. In Zen, this state is called, "don't know mind," because at that moment or during that period of mind we fully accept not knowing, we are free from knowing or having to know. I had no idea of what I would do next in my life other than visit New York and see what developed.

This state of "don't know mind" gave me permission to write to Dr. Tyson for an interview. The half-hour interview turned into an hour and six minutes. The interview beings in chapter 4 and subsequent chapters.

Please not that text in [brackets] are my notes. They explain and/or clarify the text of the interview.

Dr. Jerome Freedman

2 A SHORT HISTORY OF THE COSMOS

Cosmology is the study of the entire universe, or at least as much of the universe that we can see with our telescopes and other instruments and even the parts we can speculate about. Scientists who study cosmology are usually astrophysicists and they call themselves cosmologists.

One of their current views of the universe is the theory of the *Big Bang* which I'm sure you have heard of.

But what is the big bang? What banged? When did it happen?

Without going into too many scientific details, you will learn a little bit about the big bang, its relationship to cosmology and the importance of it for our lives here on planet Earth.

The Big Bang

Cosmologists say that the universe as we know it began about 13.7 billion years ago in a flash of energy that resulted in electromagnetic radiation and some particulate matter that was so hot that matter as we know it could not even form.

As soon as the universe came into being, it instantaneously expanded in a process known as inflation by a factor of a billion billion billion times.

After inflation, which lasted less time than you can even imagine, the universe was a seething, hot soup of particles and radiation. The universe was so dense that the radiation could not escape.

380,000 years after the new universe was expanding and cooling, it had cooled enough so that radiation could escape and this radiation is still detectable today! This is called the "cosmic microwave background radiation."

It was about this time that electrons combined with protons and neutrons to form hydrogen and helium atoms.

Initially, 25% of all atoms in the universe were helium, and a small fraction was lithium. There was no carbon, oxygen, nitrogen, or other elements essential for life.

A billion years after the big bang, gravity brought together the helium and hydrogen gases and caused them to coalesce into giant dust clouds. When the clouds became dense enough, they ignited and became the first stars.

Due to the gravitational attraction of the stars for each other, they eventually formed galaxies. As the galaxies clustered together the first stars died and in the process, produced the heavier elements in space.

The Solar System

After many generations of such stars forming and dying, there was enough matter in this region of the universe for the solar system to form. This happened about 4.5 billion years ago.

Dust particles of all sizes coalesced together to form pebbles. Pebbles came together to form rocks. As the rocks got bigger, boulders formed. Then asteroids, from about a mile across to the size of a small town came together over millions of years. Comets also formed and, ultimately planets. One of the planets – the third one from the sun – lies in what astronomers call the *"Goldilocks Zone"* where liquid water can flourish. This is where we live today on planet Earth.

The Earth

For millions of years, the Earth was bombarded with meteors, comets, asteroids, planetesimals, and other objects and was a red hot ball of lava during the period called "the great bombardment."

When the Earth was about 50 million years old, one of these objects, which may have been the size of Mars crashed into Earth and broke out to become our Moon.

All of these collisions, coupled with the heavier, radioactive materials in the center of the Earth kept the temperature of the core high enough for the core to be a molten ball of iron and nickel. Since all of this matter in the Earth's core was rotating, the Earth developed a magnetic field. It was this magnetic field that deflected high energy particles and radiation from the Sun away from the Earth, thus providing a shield for the primordial atmosphere.

The primordial atmosphere was filled with noxious gasses like carbon dioxide, sulfur dioxide, and others. The Earth took about a billion years to cool so that water could exist on the surface and land masses, first the size of islands could form.

Life

Life in the form of microbes appeared about 3.8 billion years ago. This was the primary form life until about 560 million years ago, when simple animals evolved.

For most of those 3.2 billion years between microbes and simple animals, cyanobacteria harnessed the light of the Sun through the process of photosynthesis. They increased the concentration of oxygen in the atmosphere from virtually zero to the present day value of 20%. This caused the mass extinction of species who could not tolerate such a high concentration of oxygen.

Fish have been around for about 500 million years and amphibians for about 360 million years. Birds came about 150 million years ago and flowers about 130 million years ago.

Primates flourished about 60 million years ago and the first humanoids evolved 20 million years ago. The modern human species only appeared 200,000 years ago! We are the newcomers on the block and look what we have done to Mother Earth. That is the subject of another book!

So, we, as humans, are here only because of

- The Big Bang
- Supernovae exploded in this part of the universe creating all the heavy elements needed for life
- The solar system coalesced out of the gas and debris from the explosions
- Planet Earth in the Goldie Locks Zone where liquid water can exist
- A colossal collision between the Earth and another object about the size of Mars gave rise to our Moon, which caused the axis of rotation of the earth to tilt creating seasons, and the tidal forces which stirred up the early earth
- Cyanobacteria over a period of 3.2 billion years changed the concentration of free oxygen in the air to what it is today, allowing amphibians, reptiles, mammals, birds, flowers and primates to evolve before us
- Conditions were sufficient for primates to evolve into humanoids and homosapians
- There was enough food, oxygen, water, kindness, and wisdom to bring us to present time!

3 FOUNDATIONS OF BUDDHIST THOUGHT

Introduction

The Buddha was born with the name Siddhārtha Gautama sometime around 2,600 years ago (563 BC) in the small kingdom of the Shakya Republic in what is now known as the town of Lumbini in Napal. His father was Śuddhodana, King of the Shakya clan.

His mother, Queen Maha Maya, died during childbirth and he was raised by his mother's youngest sister, Maha Pajapati.

He was not a god, but a human being like you and me.

Siddhārtha lived a princely life in separate seasonal palaces for the first 29 years of his life. He married Yaśodharā and had a child who he named Rāhula.

Whenever he left one of the palaces, his father made sure that the village outside the palace walls was cleaned up so that Siddhārtha would not be exposed to the suffering of the outside world.

One day when he left the palace, he saw a sickly person and asked his charioteer, Channa whether something like that would happen to him. Channa told him that it happens to everyone!

On another visit outside the palace, he saw an old man struggling to walk through the village. Channa once again told him that old age happens to everyone.

On yet another excursion, he saw a corpse, and Channa explained that everyone will die.

On a fourth occasion, he saw a wandering mendicant and asked who he was. Channa told him that he was a seeker of truth and freedom from suffering.

These four outings are known as the four messengers that inspired Siddhārtha to leave the palace and find his own way to relieve his suffering and the suffering of others.

For six years, he wandered around the regions of what is today Northern India in search of liberation. He had many teachers and tried all kinds of ascetic practices. Each teacher wanted Siddhārtha to take over his group of followers and become their teacher. Every time, Siddhārtha refused.

One day, he settled down by a papilla tree, the Bodhi tree, near what is now the town of Bodh Gaya. At the age of 10, he had drifted off into a deep meditation under a tree during a traditional planting ceremony on one of Śuddhodana's estates. This became the motivation for him to sit under the Bodhi tree and achieve his awakened state. From then on, he was known as the Buddha.

After his awakening, he spent the next 45 years teaching people from all castes and all walks of life only one thing: "I teach only suffering and the transformation of suffering."

The Four Noble Truths

The foundations of Buddhist thought begin with the Four Noble Truths. These truths came to the Buddha after six years of ascetic practice when he sat alone under the Bodhi tree in Bodh Gaya, India.

The first noble truth is known as the truth of suffering. Zen Master Thich Nhat Hanh likes to refer to it as the truth of ill-being. Western scholars have finally given up on the long-held notion that Buddhism is all about suffering. In some ways they shouldn't have!

We suffer when we are physically ill, often much more than we have to. We add our thoughts and feelings of fear on top of the physical illness and this added element of suffering is what the Buddha was referring to.

He was saying that we can't avoid illness, old age, and death. We can let go of our notions about illness, old age, and death and just live life in the present moment just as it is.

He also taught that we suffer when we get what we want

as well as when we get what we don't want. Funny, huh?

It is the wanting that is the problem!

And that is the second noble truth – the noble truth of desire. "To desire what we don't have is to waste what we do have," as my wife would always say to our children as they were growing up.

This desire, craving, wanting, yearning, and longing are the bases of our suffering – our ill-being, which brings us to the third noble truth, the truth of the cessation of suffering or well-being. So to say that there is only suffering is to ignore the third and fourth noble truths.

The fourth noble truth is the path out of suffering – the Noble Eightfold Path.

The eight teachings of the eightfold path are Right View, Right Thinking, Right Speech, Right Action, Right Livelihood, Right Diligence, Right Mindfulness, and Right Concentration. All of these are interconnected and are to be practiced together – not in any specific order.

The Eightfold Path

The Eightfold Path is the fourth noble truth of the Buddha. The elements of The Eightfold Path can be characterized in three groups, which serve as the cornerstones of Buddhist practice.

The first group is ethical practice, or as Zen Master Thich Nhat Hanh likes to call it, applied ethics. This group includes
- Right Speech
- Right Action
- Right Livelihood

The second group is mental discipline and includes
- Right Diligence
- Right Mindfulness
- Right Concentration

The third group is wisdom and include

- Right View
- Right Thought

Right View

Presentation of the elements of the Eightfold Path usually begins with Right View. Right view is the understanding that we must be present in the here and in the now in order to enjoy the wonders of life. The wonders of life include our bodies, our feelings, our minds, and our connection to everything around us.

We marvel in the beauty of nature and the mysteries of life in the universe. When was the last time you really enjoyed a beautiful sunset?

How about listening to the sound of the birds? When did you last see the every changing flow of a river or the stillness of a calm lake?

Think about the clouds and the rain and how important they actually are for us to be alive. And be thankful for the trees in the forest, for without them, there would be no us.

By thinking outside of ourselves, we dis-identify from our egos and recognize the beauty that surrounds us. Perhaps it is our spouse or our children that inspire us to enjoy life. Maybe we love the outdoors and spend a lot of time biking or hiking or playing.

Zen Master Thich Nhat Hanh says,

> "Our happiness and the happiness of those around us depend on our degree of Right View. Touching reality deeply — knowing what is going on inside and outside of ourselves — is the way to liberate ourselves from the suffering that is caused by wrong perceptions. Right View develops through practicing the Eight-fold Path of enlightenment. Right View is not an ideology, a system, or even a path. It is the insight we have into the reality of life, a living insight that fills us with understanding, peace, and love."

Right Speech

The second element of the Eightfold Path is Right Speech. The practice of Right Speech begins with the cultivation of loving speech. Loving speech brings joy and happiness to ourselves and the people we love and relieves their suffering.

We practice loving speech by being truthful and only saying words that inspire love, self-confidence, joy, and hope. We only spread news about things we are certain and refrain from criticizing and condemning things that we are not sure of. We avoid saying things that cause our family or community to break. Mostly, we recognize the importance of keeping Right View in mind whenever we speak to others. In conjunction with Right Speech, we practice deep listening to really hear what others are saying and reduce their suffering.

Right Thought

Once again, with Right Thought, we recognize its connection with Right View and confine our thinking to skillful means. The latter may be easy if we live in a monastery but in normal life, this can be difficult.

Every thought we have is preceded by a perception. If we ask ourselves, "Am I sure that my perception is accurate?" If not, we can simply let the thought go.

Another thing we can do to practice Right Thought is to ask ourselves, "What am I doing?" This brings us back to the present moment and allows for unskillful thoughts to vanish.

We must also pay attention to our "habit energy," as Zen Master Thich Nhat Hanh calls it. This will help us have pure thoughts and let go of bad ones. Awakening our bodhichitta - our heart-mind, our mind of awakening will render harmful thoughts useless.

Right Action

According to Zen Master Thich Nhat Hanh, "The basis of Right Action is to do everything in mindfulness." We are encouraged to wake up mindfully, walk mindfully, eat mindfully, meet with our loved ones and friends mindfully, and do everything else mindfully.

When we engage in activism for various causes we should do so mindfully, without attachment to the results. For example, if we are concerned with global warming, we can support an organization like 350.org and do so with wisdom and compassion and without any notion of personal gain.

Right Livelihood

Choosing a vocation that will not harm sentient beings and nature is the practice of Right Livelihood. Doing so will help us realize our goal of wisdom and compassion and integrate with the other elements of the Eightfold Path.

Being mindful of the global social, political, and economic situation in the world today, we also practice mindful consumption by not supporting businesses and organizations that deprive others of their chance to live a happy and productive life.

Right Diligence

Right Diligence is the chosen way of expressing the idea of Right Effort by Thich Nhat Hanh. Diligence implies investigation into what works to keep us on the Eightfold Path. Effort implies that our energy can be spent very quickly and have nothing left for practice.

Right Diligence is associated with four practices that are conducive to liberation on the Eightfold Path. These are using skillful means to

1. prevent unwholesome thoughts from arising if

they have not arisen,
2. allowing unwholesome thoughts that have arisen to subside and grow weaker,
3. encourage wholesome thoughts that have not arisen to arise,
4. and nourish wholesome thoughts that have already arisen so that they stay present in our minds and grow stronger.

The skillful means that we employ generate a feeling of joy and interest in our practice and our desire to end ill-being.

Right Mindfulness

Right Mindfulness is the heart of Buddha's teaching. Thich Nhat Hanh says in The Heart of the Buddha's Teaching: Transforming Suffering into Peace, Joy, and Liberation, "When Right Mindfulness is present, the Four Noble Truths and the other seven elements of the Eightfold Path are also present."

Jon Kabat-Zinn's definition of mindfulness is "Mindfulness means paying attention in a particular way; on purpose, in the present moment, and nonjudgmentally." He often adds, "...as if your life depended on it – and it usually does."

Right Concentration

The purpose of Right Concentration is to develop a one-pointed mind. The mind can be focused entirely on a single object and this is known as selective concentration. The mind can also be allowed to welcome whatever comes, dwelling happily in the present moment like a clear lake reflecting everything that shine on it. This is active concentration.

Right Concentration in either case gives rise to insight of interbeing – the interconnectedness of all things.

Co-dependent Origination

One of the principal teachings of the Buddha in on dependent origination. Co-dependent origination describes the chain of causation in the cycle of rebirth.

It was taught by the Buddha, himself as way to transcend suffering and achieve nirvana.

In physics, this corresponds roughly to the principle of cause and effect. However, this can be misleading because we normally think of these as being two separate events, with cause always preceding effect. Also, we think that one cause leads to one effect.

Co-dependent origination implies that cause and effect arise together – they co-arise together and everything is a result of multiple causes and conditions.

When the causes and conditions are suitable, the sun rises, the corn grows, and the rain falls.

Just think about how many causes and conditions had to be in place for life to flourish on planet Earth.

Consider, for example, our life on Earth. We know that we have the basic needs of food, clothing and shelter. Unless we have an organic garden in our backyard, we have to purchase groceries from our neighborhood store. We take it for granted that food will be there, and we rarely think about the gratitude we owe the farmer, the trucker, the store owner, the stock person, the cashier, the car manufacturer, and the list goes on.

Similar arguments can be offered for clothing and shelter.

When we look deeply into co-dependent arising, we notice two things. First, we notice is that all things interdepend on each other. This is the nature of "interbeing."

Interbeing

Interbeing is a word coined by Zen Master Thich Nhat Hanh to express the interdependent nature of all existence.

The second thing we notice is that "this is like this because that is like that!" This is an easy way of explaining Interbeing.

Suppose you are reading a printed copy of this book instead of an electronic copy. Can you see the sunshine in the paper? Can you see the cloud in the paper? Without rain, the forest could not exist.

What about the tree that the paper came from? How about the logger who helped bring the wood to the paper mill? Can you see the operator of the paper mill and the truck that transported the paper to the stationary store where the paper was purchased? How about the printer who made the words visible on the page?

So you see, the paper has to inter-be with all these causes and conditions and many more. The paper doesn't have an independent existence! The paper is made up only of non-paper elements.

Emptiness and Non-self

Since the paper cannot exist alone, we say it is empty of a separate existence. Nothing can exist by itself. Not you. Not me. Not the cypress tree in the courtyard. Nothing. If we look deeply into our true nature, we will recognize that we, too, cannot exist by ourselves alone. We inter-are with the non-us elements and this makes it possible for us to exist.

We are empty of a separate existence, just like the paper. We have no separate self. With this insight comes the insight of Interbeing. Then you come to know that your happiness and suffering depend on the happiness and suffering of others. If you help other people suffer less, you will suffer less.

Dr. Jerome Freedman

4 THE STORIES WE TELL

Dr. Freedman: So there's a story involved. Of course, everybody has a story. Not only did I study physics in the 60s at the University of Chicago and Washington University. Nothing really clicked. And I didn't get a PhD in Physics. I eventually got one in Computer Science. And in recent years I've been studying the teachings of the Buddha.

Dr. Tyson: And you said...?

Dr. Freedman: Twenty five when I met this Vietnamese teacher named Thich Nhat Hanh.

He was pretty well known for having made fantastic efforts to stop the war in Vietnam and he was nominated by Dr. Martin Luther King for the Nobel Peace Prize. But he did not get it but it doesn't matter. He doesn't care. I care but he doesn't.

And I struck up a friendship with him, learned a lot from him over the years.

And about four years ago, five years ago, it might have been, I caught one of the episodes of The Universe. And you were on it and also Alex Filippenko was on it. And I was intrigued.

And I also saw Origins and I was very much intrigued because I'm interested in the connection between life and the universe and between the teachings of the Buddha and the teachings of cosmology.

Because I think there is a lot of overlap. Even though the teachings of the Buddha are not quantitative and they don't really ask questions that don't have an easy answer because it's more about practice.

It's more about what can you use to be a better person to live a compassionate life to understand what people have to say and to enjoy your life fully in the present moment.

So that started me off on a little crusade to investigate the nature of the universe from the point of view of cosmology. And then tie in it to include Buddhist stuff.

So about two years ago I started emailing Alex Filippenko and he emailed me back and I watched one of his Learning Company videos.

Dr Tyson: Teaching Company.

Dr. Freedman: Teaching Company yeah.

On one of them, he was talking about tennis. Well I'm an avid tennis player. I play like three or four times a week. So I figured uh, a connection.

But then later he talked about Japanese food. And my favorite Japanese restaurant is Berkeley where you know he's a professor there.

So I wrote him again and he gave a lecture last year in Los Altos at one of the colleges. And I went and I met him and we talked for 10 minutes.

And then I went to SETIcon in August with my friend, and we had a great time.

SETIcon was in San Jose. Do you ever go to those?

Dr. Tyson: No.

Dr. Freedman: So I've kind of been stalking Alex Filippenko a little bit to see if I can entice him into a Japanese meal or for getting him in a game of tennis, but that hasn't come to fruition yet.

I'm still very fascinated by the connections that I have not discovered, that have always been there that I've thought about of times of relating this stuff through Buddhist thought and cosmology. So you come in as a cosmologist or astronomer, however, I'm not sure how you call it.

Dr. Tyson: Astrophysicist.

Dr. Freedman: Astrophysicist. Yeah, that's what you say on TV. And I thought I might introduce you to some of the thoughts that I've been having to see what you have to say about them.

Dr. Tyson: Okay. And towards what end is all of this?

Dr. Freedman: Oh, see I maintain a website on Buddhist thought called Mountain Sangha, which is a group of people that I meditate with on a regular basis. And I'm trying to write an article about the relationship between Buddha thought and cosmology and gathering material.

Dr. Tyson: An article for the website.

Dr. Freedman: For the website. And there's also quite a number of Buddhist journals that I would try to publish it in before I'd make it publicly available on the website. Because I need to see the subject better.

Many Buddhists would be interested in if they knew that there was something there.

I don't think there are too many physicists who became Buddhists and vice versa.

Dr. Tyson: I know a couple.

Dr. Freedman: Do you?

Dr. Tyson: Yes, I know at least one.

Someone in fact with my same last name, Tyson, Tony Tyson.

Dr. Freedman: Oh really. And where is he?

I don't know where he's lately. His career was spent in Bell Laboratories. I think he went to California.

Dr. Freedman: Tony Tyson.

Dr. Tyson: Tony Tyson yeah. There are only two Tyson's in the US.

5 ORIGINS AND THE GOLDILOCKS ZONE

Dr. Freedman: So I guess a good starting point is to just consider Origins.

Now for the lives developed on this Earth and what happened is kind of miraculous in a lot of ways until we really unraveled all of the secrets of what it takes to create life. And that, the first principles there...

Dr. Tyson: So the people who say what they don't understand, those will call that a miracle are not likely to be the ones to discover how it works.

Dr. Freedman: Yes.

Dr. Tyson: I just want to alert you to this.

Dr. Freedman: I agree with that.

Dr. Tyson: So to say it's a miracle, you don't understand it. I'm saying it's an interesting problem that we're working on. I don't think it was a miracle.

Dr. Freedman: I like that way of thinking.

Dr. Tyson: Just letting you know.

Dr. Freedman: Yeah, that's perfect. That's exactly the kind of insight that I'm looking to get out of this interview.

So with life you have the mechanisms for survival and the mechanisms for reproduction. And then you have human beings who came along and they have something else.

There seems to be, we seem to have something else going for us because we recognize happiness and suffering. And as the Dalai Lama says, everybody wants to avoid suffering and to have happiness.

And that's another motivating factor in life.

Dr. Tyson: So that is true in non-human life forms as well.

Dr. Freedman: Yeah, probably you're right, yeah.

Dr. Tyson: We're most certainly not unique among animals. And any animal that feels pain wants to avoid suffering. So if you attempt to inflict pain on an animal they will resist you. That's the avoidance of suffering.

And most mammals particularly in their infancy enjoy playing. So any understanding of animal behavior would make a value on happiness. So to say humans come along and you have something different is a culturally biased assessment of the role of humans on this planet.
[A recent article on MyFDL (my.firedoglake.com) indicated that India recognizes dolphins as non-human persons whose rights to life and liberty must be respected.]

Dr. Freedman: Okay, I buy that.

Dr. Tyson: And it's rampant in much literature where there's an attempt to distinguish humans from other animals, where if we go far enough back in time earth from other places in the universe. Of course, earth is doing all it can to try and kill us. So we sweep that under the rug and say, "Oh, earth is a haven."

So there's a reality check that is not often enough for us to bear on many philosophies that like to think of humans as something apart or different from the rest of the animal kingdom.

Dr. Freedman: Yes, I totally agree.

Dr. Tyson: I interrupt. Sorry.

Dr. Freedman: No, that's good. I like it. That's the important thing. It's like people who rattle on and on and on without letting somebody interrupt miss a lot...

Dr. Tyson: Well this is New York so many interruptions are welcome, actually.

Dr. Freedman: So when you think about life and think about Origins like your wonderful presentation. And then you wonder about the place of the Earth in the universe. And we're in this Goldilocks zone between being too close to a star, it will burn up and if too far away from the star, for there to be no water, no ice or maybe nothing to support to life.

Dr. Tyson: It's a liquid water issue.

Dr. Freedman: Yes.

Dr. Tyson: So life as we know it, requires liquid water. And the Goldilocks zone operates on that premise. Some other life that we would come about would require liquid water as well.

So that involves a bias that we're self-aware of but there are other zones for example ammonia has a different Goldilocks zone than water does. And it's a fluid and it can carry nutrients.

And so one could ask if you open your mind to possibilities beyond what we are, perhaps there could be life that thrives on ammonia.

Dr. Freedman: Or liquid methane.

Dr. Tyson: So liquid water, liquid methane, exactly. Liquid methane, it moons of Titan – one of the moons of Saturn. So you can say we're in the Goldilocks zone and we can define all the things that make that special for us. But it could be in fact hostile for some other creature.

Dr. Freedman: For sure. Yes the Andromeda Strain movie comes to mind.

Dr. Tyson: Sure.

6 EXOPLANETS

Dr. Freedman: So what are the latest discoveries that have to do with exoplanets right now?

Dr. Tyson: I was just building the inventory for the planets right now. And it's up in the thousands, through a dedicated telescope which is...

Dr. Freedman: Is that the Fermi telescope?

Dr. Tyson: No, no. no. Kepler. And it's measuring what we call transits.
   You might almost call it an eclipse but it's not big enough to block the light completely of the host stars. So a planet moving in front, you see a different in the light of the planet as you track of how bright it is over time.
   And you can deduce quite a bit about the mass of the planet, what its orbit is, whether it is eccentric; is it in a Goldilocks zone of the host star.
   And so we're building that inventory, learning about whether our solar system is common or uncommon for whether Goldilocks planets are common or uncommon, this sort of thing. And so they're coming in daily, monthly, weekly.

Dr. Freedman: Yes, I know. And there are guys at Berkeley that are doing it. What are your thoughts about life in another solar system?

Dr. Tyson: You mean what do I think of the likelihood?

Dr. Freedman: Yes.

Dr. Tyson: Certainty. Oh yeah. We have no data, no evidence and so the confidence I have in that statement derives from secondary arguments, not primary arguments.

The primary argument was we found some life.

The secondary arguments are life on earth is made of the most common ingredients in the universe.

Dr. Freedman: Carbon, oxygen...

Dr. Tyson: The most common ingredients of the universe. There is nothing rare about it.

And we are chemically based on carbon, which is the most chemically fertile element in the universe.

So you combine the abundance of these element and their chemical fertility and given that the most complex form of chemistry we know we call biology, you stand the best chance to achieve complex chemistry, if you base your chemistry on carbon.

So not only that life on earth began almost as soon as it possibly could – a very short in time. Around 200 million years it turns out.

We used to think it was longer than that and we were perfectly happy calling that short. We thought that maybe 600 million years or like a billion years. But as fossil evidence goes farther and farther back in time and as we learn about the formation of the earth, how it had been even hostile to the formation of complex chemistry.

Those two are pinching the time interval over which inanimate molecules would have become life. And that interval is a couple hundred million years. Which is a snap of the history of the Earth.

Dr. Freedman: And the history of the universe.

Dr. Tyson: And the history of the universe. Right.

Dr. Freedman: So this is all, this stuff that, am really into this subject matter...

Dr. Tyson: What I find curious is there's a risk of exploring where there's overlap or resonance for...

Dr. Freedman: Resonance.

Dr. Tyson: Resonance is probably a better term. And what does it mean, if I can ask you a question. What does it mean to look for a resonance? Suppose half the things I told you don't resonate at all and the other half do, do you count that as a hit and you say there is resonance between Buddha's teachings and cosmology?

Or are you going to say no, half the time they get it wrong or it's not resonates so we just give up?

Are you going to pick and choose what does resonate and say it resonates at the end? Because it's very easy to then not reflect on the things that don't resonate and then sift through what does and present that.

And then give the misleading impression that Buddhism is perfectly aligned with modern cosmology. By the selection of data that you invoke.

Dr. Freedman: I am thrilled with your question because it really leads me into one of the topics on my cheat sheet. And that topic is...

Dr. Tyson: Because you want to see if it resonates, you should come to me with questions that derive from Buddhism. And I will tell you whether or not that fits with cosmology.

Dr. Freedman: Well let's talk about it.

Dr. Tyson: Because that's a much stronger way to approach this then to have me tell you all of my cosmology and then you go home and pick and choose what fits the philosophies of Buddhism.

Dr. Freedman: Well, I was going to start with this particular concept. One of the principles of Buddhist thought is that when things are...

Dr. Tyson: Oh just so you know, I have the Reader's Digest knowledge of Buddhism so when in doubt, assume I don't know what you're going to tell me.

Dr. Freedman: Fine, I was not planning to assume anything about Buddhism...

Dr. Tyson: Because I know a couple of things but not enough to engage in your conversation.

Dr. Freedman: Yeah, I am that way with cosmology.

7 CO-DEPENDENT ARISING, INTERBEING, AND IMPERMANENCE

Dr. Freedman: So the Buddhist principle that I wanted to bring up first was this idea of co-dependent arising. In other words, when causes and conditions are such then something will arise. And when those conditions are no longer present, something will go away.

So the formation of life on earth had that kind of environment where causes and the condition were such that life could evolve. And we're continuing to see that happen except in our generation what's happening is we're ignoring this concept.

And that one leads to another concept that everything in the universe is interconnected. The thing about the butterfly in South America causing reverberations on the furthest galaxy.

Interconnectedness and dependent co-arising of nature are two principle ideas in Buddhism that really mesh nicely with cosmology.

The third one is what you might call impermanence where everything changes. Nothing stays the same.

Planets evolve in some way or the other. Stars are formed by gases coming together and converging and their reaction starting up. And all kinds of events demonstrate this principle.

And Buddha thought that that was a fundamental reality of life.

Although it might seem quite obvious to anybody who's thinking about it, it's really an amazing concept because with this idea of interconnectedness, the guys in Washington DC who are only interested in their self-aggrandizement and becoming leaders and having power, don't take into account that whatever they're doing has an effect on other people. They just do it for themselves.

And the idea of interdependence means that when I make a job for someone in India, I am supposedly elevating his quality of life. And I don't necessarily have to reduce my quality life but it would be helpful to do so.

Dr. Tyson: For the economics of it.

Dr. Freedman: Yeah, for the economics of it. And also for the relationship between countries is based on we are one.

But each country thinks of itself as an independent entity. But what would happen if we decided we would cut off relationships with every country? We'd be in a really bad place. And we would recognize how interconnected we are with all the other countries of the world.

And I think that this interdependence concept comes into play in cosmology because, for example, the web of the cosmic background radiation produce a web of galaxies and where dark matter and dark energy evolve... I'll let you explain that part of it.

But I see that as interconnection, interconnected, interbeing as my teacher would say. And I think that when one becomes knowledgeable about interbeing, one is able to be more compassionate, more understanding, more able to form relationships, more able to connect and more able to make a difference in the world.

So what do you think about the concept of interdependence as I explained it?

Dr. Tyson: I would say in modern times, (I need to define modern as in the last century). What you said recent times was 25 years. In modern times, we have come to learn about ecology. I'll use a single word but more specifically the interdependence of life, animal life, plant life, water supply, atmosphere.

It's a system; systems engineering is all about interconnectivity and parts that create one functioning whole.

Cosmology and Buddhist Thought

These concepts emerged as 20th Century revelations about the world that we live in.

So you can say that Buddha, Buddhist teachings knew this from the beginning. However, if you go before the 20th Century, in mid-19th Century, so go earlier than that. What you did had very little consequence outside of your zone, outside of your...

Dr. Freedman: Your quote, "system."

Dr. Tyson: Outside of your [zone], people were far enough apart and were not so aggressive on the environment that their behavior would affect someplace else. So that in fact there was not this deep interconnectedness of it all. Because there was a susceptible...

Dr. Freedman: But was it there...

Dr. Tyson: Not in any meaningful way.

Dr. Freedman: Right. It was there imperceptibly.

Dr. Tyson: Okay but the butterfly effect was an overplayed media account of an attempt to bring the concept of chaos to the public. So I'm just simply saying, that when you want to talk about interconnectedness today, the fact that we fly airplanes from continent to continent and move goods and services from continent to continent and insects, vermin, whatever, ride ships from one place to another.

And the fact that we change gases in the atmosphere here that then circulate around the globe, to say that we are interconnected today with the same fervor as that we were interconnected a thousand years ago, is just misusing the word. It's using the word in such a way so that Buddha was not wrong.

All I'm saying is, back then interconnectedness had no meaningful consequence to anything. It required major travel and environmental disruption that came about in the era of technology, in the industrial revolution essentially.

And so sure, you can be say it's all been connected the whole time but there's no contest in these two cases. And if you don't want to distinguish those two cases, then it's hard to have a conversation about what it means to be interconnected.

So for example, in my concluding words from The Universe series, when I say we're connected, I'm using the word connected differently from how you meant it and how you just described it.

And the way I'm using it is the carbon that is in your body is the same carbon that is across the universe. And it has similar points of origin – origin in the centers of stars. So that shared identity, what I call the connectivity.

The way you use the word connectivity just now is that one has an influence over the other. And that's just simply not the case in the universe. Your carbon atoms are not affecting the carbon atoms across the universe. They are not connected in the way we speak of connectivity in the global ecology for example. So we need another word.

Maybe I should have used a different word. I could have said there is a shared heritage. Otherwise I could have said it. But what happens on earth and in our solar system. There is in our galaxy, we feel the gravity of another galaxy. We are going to collide with the Andromeda galaxy. That's scheduled to happen after the sun dies.

So you can say yeah we're still all connected. But it's kind of irrelevant because we'll be vaporized. You know it's kind of irrelevant insight into the universe to say we're connected because we're gravitationally bound with another galaxy who we're about to collide.

There's a horizon of the universe that's expanding. Beyond that horizon, we don't even feel each others' gravity.

It's beyond any accessibility to us. So everything is not

connected in the same way that I destroy this air and it effects the ecosystem on the other side of the earth.

There's a huge spectrum of connectivity, some of which is just simply irrelevant to anything that matters to anybody at any time, at any place. You know the things that are really relevant that affect the quality of the air you breathe that will determine what kind of life you lead. Whether it will be a healthy life or a sickly life. Because we're interfering with the water supply and air supply, your climate, whatever else. So cosmologically speaking, the fact that we share the same ingredients doesn't mean we're causally connected in any fundamental way.

By the way, you see the light that comes from them because they were connected that way. That emitted light, that's how we know it's a carbon. Because carbon are emitted to absorb into the spectrum. But it crossed the galaxy and entered our detectors. So there's cause and effect there. Okay, do you want to say it's all connected because of that one fact?

Dr. Freedman: Well it's not the only fact. It's not a single fact.

Dr. Tyson: Well I'm just saying there's a spectrum. And so the word is rapidly loses its utility. If you were going to put all of this variation of cause and effect under the same word, then there's no way to test the concept if anything works for it. It's the old saying, if it explains everything, then it explains nothing.

That was part of the problem with chaos. If you can say everything comes from chaos, but in fact you can't really test that. You can't test that everything comes from chaos. Because whatever is the result, it's good with you. It's some chaotic path led to that storm. Well a different chaotic path led to some other storm but that ultimately wasn't the storm we got. We got this storm. Well that other storm comes in, see that chaos. You got this storm so you have chaos. It's not useful.

It becomes metaphysics at that level. And metaphysics has never been accused of being useful. But it provides great conversation with a beer at a pub.

Dr. Freedman: I have one example of interconnectedness, of interbeing ...

Dr. Tyson: Hold on, I have something to say about the ending of what you say that everything is ephemeral – is that the word used?

Dr. Freedman: Oh I used impermanent. Impermanent.

Dr. Tyson: I have a quick comment about that.

8 TIME SCALES

Dr. Tyson: I love comparing time scales of things. That's a favorite past time of the astrophysicist.

In fact, there's a very cute, and forgive me for not remembering the name of it, once you look it up it will be more effort.

There was a movie short that was, that I had either won an award or was the runner up for the award for the best movie short of the year. [Das Rad]

Dr. Freedman: In what year? That was many years ago?

Dr. Tyson: In the decade, the last ten years.

Dr. Freedman: And the powers of ten?

Dr. Tyson: No, no, no. It's just a German short. There were two rocks having a conversation with each other. And one of them asked, "Do lichens irritate you?"

And he goes, "Yeah, they're kind of itchy. And they on grow on my back. And the try to reshape you."

And he said these two rocks are talking about lichens. But lichens take a long time to show up on a rock and the conditions have to be right and all that.

Dr. Freedman: Ah, causes and conditions.

Dr. Tyson: And you look at the environment that the rocks are in, you don't quite understand what's going on because the sky is kind of pulsing in a weird sort of way. You don't really understand it.

And then...

Dr. Freedman: And then there's this gravitational wave from the big bang...

Dr. Tyson: No, no, no. And so it's, then you hear this slowing down sound. Like rrrrrr, like that. And then you see the sun in the sky.

And the two cavemen walk up to the rock. And then there is a rock sitting on top on the rock and he grabs it and looks at it and then walks off.

And only then you figure out what's going on. The sun goes around faster and faster and faster, then it's brrr, and then the rocks continue to talk to each other.

So it's like a billion years in the life of a rock shrunk down to these couple of minutes. And it was, so the rocks are observing all of these things going on but there are selected moments where they slow things down enough so you can see something else take place.

So there's a rock and you have the mayfly that lives a day. These are intriguing to me that there are all these time scales. There is a time scale that is the mother of all time scales and that's the decay of the proton, if it decays at all. It's 10^{30} years is the latest calculation. Or 10^{32}. That is 20 orders of magnitude longer than the current age of the universe.

Now do you want to include that as everything, what's the word we used again? If it dies everything...

Dr. Freedman: It is the impermanence.

Dr. Tyson: Impermanence. So even a proton decays. But that's the most stable known particle. So you can say Buddha had it, even there. And like okay. But then once again, it explains everything. And therefore leaves us with nothing. So it's not, it's really not a scientifically useful concept.

Dr. Freedman: It's not scientific. I never claimed...

Dr. Tyson: But you will, if you are because you're going to say these scientific points come back into the philosophy and there's resonance.

Whereas I submit to you that given a range of those facts it's not some deep scientific principle. It's not a, it could be something different if you, if Buddha said, he wouldn't say this but I'm making this up. If all the universes filled with energy, and that energy becomes more and more dilute, and we measure this as we feel cold. All right you can map that statement directly to the cosmic microwave background. That's kind of cool. That would be kind of cool.

Another thing, it said universe will last a long time and the Buddhist estimates of the age of the universe, are the only ones that were really long compared to all of the other estimates that came out of the east and the west. And it borders a billion. But it's off by orders of magnitude. So it was kind of cool that it was long because no one, nobody was thinking long at the time.

Dr. Freedman: And no basis for it.

Dr. Tyson: No way to even wrap their brain around it. So kind of cool that it was long. But it's off by multiple orders of magnitude. Not that science hasn't been off but multiple orders of magnitude. But we have self-correcting mechanisms and we discard it when it's wrong. And we don't return to it because it's not useful. And we get the one that's right. And then we put that into effect.
[A side note about how meditation fits into this statement: Buddha's insights into the nature of life and the universe came from his direct experience in meditation. Many followers, both men and women, over the ages have repeated his "experiments" by following his methods. They have tested his insights and found them to be true for themselves as well.

[The Buddha consistently taught that we should not go by what has been acquired by oral tradition. Neither should we be convinced by knowledge obtained from a teacher to a disciple. We must not be certain that what is written in scripture is the absolute truth, nor should reason or bias be the basis of or knowledge. We shouldn't take the advice of an expert on the matter, nor should we take even the word of the Buddha.]

[When we know for ourselves that something is good, wholesome, beneficial, and true then we can take what we experienced as something to embrace.]

[The Buddha said, "Don't blindly believe what I say. Don't believe me because others convince you of my words. Don't believe anything you see, read, or hear from others, whether of authority, religious teachers or texts. Don't rely on logic alone, nor speculation. Don't infer or be deceived by appearances."]

["Do not give up your authority and follow blindly the will of others. This way will lead to only delusion."

["Find out for yourself what is truth, what is real. Discover that there are virtuous things and there are non-virtuous things. Once you have discovered for yourself give up the bad and embrace the good."]

Dr. Freedman: The process of meditation and learning about Buddhist thought is the same kind of method. When something doesn't work, you abandon it.

Dr. Tyson: Did you catch the exhibit at the Rubin Museum?

Dr. Freedman: Yes. That was one of the first things I wanted to do.

Dr. Tyson: Good, Okay. I had one of my books in that exhibit and we produced, and we did the exhibit video of The Known Universe. Which that was a nine million hits on YouTube – the first viral video. It was a zoom out from Tibet basically all the way to the end of the universe.

Dr. Freedman: Cool, I've got to see that.

Dr. Tyson: Yeah, The Known Universe. It's six minutes and 30 seconds.

Dr. Freedman: Oh great. So I think I lost my train of thought.

Dr. Tyson: Because I keep interrupting you, sorry. The impermanence and the interconnectedness.

Dr. Freedman: Right.

Dr. Tyson: I was just addressing those.

Dr. Freedman: Right, you did a great job. I enjoyed it very much.

9 BUDDHIST THOUGHT AND THE SELF

Dr. Freedman: Let's see, in Buddhist thought, the first order of learning is your own self-realization. The second order of learning is from your "guru" if you have one – your teacher. And the third order...

Dr. Tyson: Is sensei used in that?

Dr. Freedman: Sensei is more used in martial arts.

[Sensei actually means, "a person born before another." In general usage, it means "teacher." So Dr. Tyson was right!]

Dr. Tyson: I didn't know that. That is cool. Okay. But they're still a teacher role.

Dr. Freedman: Yeah. They're definitely a teacher role.

Dr. Tyson: Thanks for that.

Dr. Freedman: And in Tibetan Buddhism they are called lamas and rimpoches depending on whether they have a past life or not which is an interesting thing to explore at some point, but it's not one of my primary things I want to explore.
    The teacher is your second line of knowledge and the scripture is your third line.
    And scripture hardly counts at all in Buddhism. Whereas in most other religions, it's the word that counts. And your own experience...

Dr. Tyson: In all the others.

Dr. Freedman: Yeah, in all the others.

Dr. Tyson: They trump anything you experience.

Dr. Freedman: Yes. So I think that was the main attraction to Buddhism for me was I have my own experience of what I think is right and what is wrong. And my own experience of how I can be a more compassionate, loving person. And I try to live by these it's called sila or ethics, by these principles.

So okay so then the next idea that really is important to explore is the nature of the self. And according to Buddhists thought and the self is empty. I am empty of an independent existence because I am made of everything that is not me.

This piece of paper, can you see the sunshine in this paper? That's how Thich Nhat Hanh explains it. And he says, if you can see the sunshine in this paper, you can see that it is empty of independent existence.

Because without the sun to make the trees, to make the loggers who make the paper mill, etc. all the interconnections that get the paper into my hands, there's no independent existence. So this leads to an interesting idea that ...

Dr. Tyson: Give me an example of what an independent existence would be.

Dr. Freedman: I can't. I don't know how to give you an example. Well I can give you an example. If you take an isolated bubble chamber that has data that you extract out of it, right. It's an isolated system. I guess that might have...

Dr. Tyson: The best experiments are designed in just that way.

Dr. Freedman: Yeah. The best experiments are designed to be...you know when you get down to quantum theory, you can't really tell the position and velocity of an electron no matter how hard you try.

Dr. Tyson: Right, so you answer all the questions that you can.

Dr. Freedman: Yes, exactly. So the idea of not, of emptiness means that a person or a piece of paper, a desk, a lamp, they don't have an independent existence.

Dr. Tyson: But what is the meaning... what is the value of having that perspective?

Dr. Freedman: The value is this.
   Here we are, you and I sitting. I think I probably know much more about you than you know about me. But I really don't know that much more about you and I just admire all of the shows that I've seen you do.
   And I felt that I wanted to speak to you and I had to see what happened. So now there's sort of a transference in which we have a communication going.
   But if you take the senators and the congressmen they have their own ideas that they are separate from everybody else and they only work for the betterment of themselves. You have a problem.
   And I'm saying that the understanding of not having an independent self or being empty of an independent self, gives people a perspective in which they're more cooperative, they love to share.

Dr. Tyson: So it puts a behavior on them.

Dr. Freedman: It does. I had a little difficulty with this one because I know that science requires isolated systems.

Dr. Tyson: Isolated as its...

Dr. Freedman: As it's possible.

Dr. Tyson: The more isolated it is the more reliable the result is.

Dr. Freedman: Right. When you try to put together multiple systems like the ecology we're referring to, you have events of a lot of trouble. Because mathematically it's virtually intractable.

But the computer simulations can figure out how the microwave background radiation maps the locations and the strengths of the streams of galaxies in the inter web there. I'm so impressed with that work. But they can't map the weather for 30 minutes.

Dr. Tyson: Five days

Dr. Freedman: Five days. I thought that was going to be thunder storms. I looked on the internet before I left Tampa and said oh no, there's going to be thunder storms. Look at the weather; I must have brought the sunshine in California and Florida.

Dr. Tyson: Well there might have been. That's where it was pretty local. And the first principle is what did you say?

10 NIRVANA

Dr. Freedman: The first principle was the principle that when causes and conditions are right, things manifest then. [If] the causes and conditions aren't there, whatever was supposed to manifest won't. It's like the storm did not manifest.

Dr. Tyson: Right and what's the value of that?

Dr. Freedman: The value of that has to do with deeply understanding changes in your life that can have consequences. I'll tell you...it actually connects to the next concept really. And you've heard of the concept of nirvana, I'm sure. And what is your understanding?

Dr. Tyson: I would say a state of mind or state of body entering in a heightened peak sense of happiness.

Dr. Freedman: Pretty much close.

Dr. Tyson: Is it mind or body or both?

Dr. Freedman: It's totally depends on whether it's mind or body or both! It can be all of the above. Or one or the other in itself in any combination. I guess.
My understanding is it's very simple and it's really being present – that life exists in the present moment only.
And that nature has this tendency to be that way. But we don't for some reason. We can fight things and therefore lose our composure, our serenity, and our equanimity.

Dr. Tyson: You say fight things. Like what?

Dr. Freedman: Well we have desire that …

Dr. Tyson: Flexible but what are we fighting about?

Dr. Freedman: We're fighting the fact that our wants, our desires are not being satisfied. And we're fighting them.

Dr. Tyson: Fighting. I still don't know what we're fighting. If you have unfulfilled desires, what are you fighting?

Dr. Freedman: You're fighting to fulfill your desire.

Dr. Tyson: So why is it a fight? I desire to have a milkshake. I really like milkshakes because it's malt in it. But it's like two miles away and I don't have a way to get there. Maybe I'll get there next week when I drive by it. So I postpone it.

Dr. Freedman: Yes.

Dr. Tyson: So what am I fighting?

Dr. Freedman: Well that's not a fight there. You've come to…

Dr. Tyson: But it's a desire I have.

Dr. Freedman: But there are all kinds of desires at all levels.

Dr. Tyson: Right, I'm just trying to get concrete example of what a desire unfulfilled that you're fighting for. I just don't know what that really means.

Dr. Freedman: Okay, let me see if I can…

Dr. Tyson: It's not a given to me that it's an unfulfilled desired leads to a fight.

Dr. Freedman: Well let's put it this way. It could be, it could lead to a struggle.

Dr. Tyson: For sure. I want to ultimately kill for it.

Dr. Freedman: Yeah right. The suffering. And to causing other people to suffer. And it is in the context of ...

Dr. Tyson: Okay, so it's not the generalized statement, it's specifically examples of desires gone unchecked.

Dr. Freedman: Yes.

Dr. Tyson: People that turn violent or disrespectful or abusive.

11 ECOLOGY

Dr. Freedman: Very good. You put it, well. Very clear thinking. Now for me...

Dr. Tyson: By the way, you've got to look at organizations in total. Humans are not the worst force ever to be set loose onto the ecology of the earth. That took place two and a half billion years ago when the cyanobacteria of the oceans slowly but systematically and irreversibly over time, converted to carbon dioxide atmosphere to an oxygen atmosphere.

And all the surface creatures that die in the presence of oxygen died. And entire new waves of life arose.

Thriving in the oxygen atmosphere and all the anaerobic creatures – they either died or they went subsurface.

And so if you ever go to a beach and you dig through the sand, below a certain distance the sand changes color almost abruptly. That's a place where no oxygen reaches it and the color comes about from the microbes that thrive anaerobically.

Dr. Freedman: I've done that.

Dr. Tyson: Yeah that's right. And the same is true at some parts of the bottom of the ocean. If there is no oxygen circulation and no oceanic currents, the water is then purple. There is some other ecosystem that is not oxygen generated. And so those bacteria completely transform the world. And far more than we ever can or will.

Dr. Freedman: I think I remember that from Origins.

Dr. Tyson: Yeah, it would be in Origins that's right.

Dr. Freedman: So it was those mounds in Australia.

Dr. Tyson: Yeah, as a record of the...

Dr. Freedman: Of the cyanobacteria.

Dr. Tyson: Right, right.

Dr. Freedman: Yeah, that's an interesting observation.

Dr. Tyson: And so I'm saying, so humans are uniquely guilty for wanting to change their environment to suit their needs. Of course, beavers do that as well. We somehow say it's all okay for them but not okay for us.

I tend to look a little more holistically at things that any animal is no longer alter its environment to serve its own needs. They all do it. What does an ant do? I want to dig in the earth and pull grains out and create them into a warren of chambers and underneath the soil for their own purposes.

Dr. Freedman: But they cooperate with each other.

Dr. Tyson: No, but not with other ant colonies.

So a lion's den does then doesn't cooperate with other lion's den.

So in other words, I'm more absolving, is that the right word? I'm more forgiving of human behavior than many other people are, when I compare it to the behavior of other creatures that in their own attempt to live do whatever they do and whatever they can to, with the environment to each other in order to survive.

12 CONSCIOUSNESS

Dr. Freedman: So yeah, well that was the first thing we talked about was survival. And the interesting thing about the cyanobacteria, the question I have about them is, "Were they conscious of that? Was there any consciousness there?"

Dr. Tyson: I'm not going to judge whether our definition of consciousness which of course is still generating huge literature on people trying to figure it out. Which is the evidence is that we have no clue what it is. And usually when something remains that intractable, it tells you that it doesn't really exist as we have been attempting to define it at all.

And maybe it's simply the wrong question it's like asking what flavor of cheese is the moon made out of. And the moon is not made out of cheese at all.

But you invest in energy trying to decided what kind of cheese they want. If we are trying to find consciousness and often it's done in ways that are tries to distinguish us from other animals.

It may be at the end of the day there's no such thing as what we call consciousness. It's something else. And we haven't been asking the right questions.

It's like I remain open to that fact given that was a retractable question of consciousness has been.

But anyone who's owned a pet knows they're completely conscious of what's going on and they don't like try to trick you. And cats and dogs and horses and those domestic animals where we give them some kind of freedom, only cows, I don't know if cows are kind of connive behind your back.

So I can probably pose an argument to say that this cyanobacteria were not conscious. It requires a map of neurons to lead to complex thoughts such as what we have in our brain.

Of course we don't have the largest brain. By far not the largest brain. And it's not even the largest brain relative to our body weight. That was put forth early on, yet another attempt to distinguish us from the rest of the animal kingdom.

And then we found that there are, I'm sorry we do have the heaviest brain relative to the body weight but if you made a line, if that was your measure of things, and you put all animals on the scale, then the some goldfish are higher on that scale than dogs are.

There is some stuff that you would not want to be true if this in fact were some deep other understanding of intellect and consciousness. So probably it wasn't conscious. It's just its waste product. That's all it is. It's waste products killed off all anaerobic life forms.

Dr. Freedman: The fact that the oxygen is in the air, has allowed mammals to develop.

Dr. Tyson: That's correct. Entirely. Like all creatures that thrive on oxygen which is quite potent. And it is a potent source of chemical energy that you can oxidize. Oxidizing you can reduce chemicals. There is a lot of energy contained in that. So change transformed their world. Interesting.

Dr. Freedman: And we're transformed back to carbon dioxide! We need some cyanobacteria to help us out.

Dr. Tyson: There you go!

So I look at the universe as a whole. Hard not to as an astrophysicist.

And I see the, like I said the connectivity in the way that I used the word and I'm sorry if it was misleading where I use it. It's not my intent.

But for me that shared genetic, that shared atomic, that

shared molecular heritage allows me to feel a part of the universe in a way that might not have otherwise empowered me to do so, short of reading someone's philosophies that are not based on the empirical discovery. And I intend to be more empirically driven in how I think about the physical world.

We provide anchors to support ideas and arguments that don't otherwise exist. And anchors that can transform a metaphysical conversation to a physical one. And then the actionable statements.

So that's how I come at it. But if you want everything to decay, you got the proton. Once sextillion time the age of the current universe, the current theories say that it will decay and then you've got everything. Chock it up as Buddhistic if you like!

Dr. Jerome Freedman

13 RELIGION

Dr. Freedman: See what I'm trying to do is...

Dr. Tyson: Let me say it in another way. Here's something you might use.

Let's not distract ourselves as you open your conversation with the absence of quantitative value to these comments or to these teachings.

Let's not even go there. Let's just recognize that there are statements about the physical world that are made that are less conflicting with what science has revealed about the natural world than other philosophies, particularly religious philosophies – less conflicting.

You will explore this further and decide whether or not it is resonates. The most I'll give you is that it is less conflicting to say the universe is 100 billion years old and to say it's 6,000.

The idea is that it's really, really old. It got there. It landed there. Or whatever was the number. But it's way more than the current age of the universe.

So to factor all of this together, one might ask an interesting question. You might write another article on this; you should, I think.

If it's less conflicting with the actual nature of the universe and religious teachings of Protestants and Catholics and Quantic scholars and if that conflicts then throughout history you would have perhaps found more Buddhists represented among the greatest scientists there ever was then people from other deeply held philosophy.

In sense that they would have dig out of those of philosophies that were directly conflicting with what they were trying to discover about the universe and never hold them back, cage them, handcuff them, leave them struggling just to reconcile the two. Whereas in Buddhism there's not a challenge to reconcile it because it's not really preventing them from having those thoughts.

Dr. Freedman: That's a real good point of view.

Dr. Tyson: The problem is that's folded together with the discovery of science as an enterprise anyway, which is the western thing and not an eastern thing.

Dr. Freedman: Oh that's not necessarily so.

Dr. Tyson: Yes, science has practiced that it's 100% so.

Dr. Freedman: Oh it is, okay.

Dr. Tyson: There's no ...

Dr. Freedman: There's known empirical stuff.

Dr. Tyson: But that's what science is. That's my point. It's not science if you can't make a testable prediction. It's metaphysics. So there are tremendous metaphysics traditions in the far east. None of it leads to discovery.

Dr. Freedman: Well one of the things that's going on...

Dr. Tyson: My point is that since science was invented in the west. Had it been invented in the east, had it been invented in both places at the same time, I bet you it would have risen faster in the east then it did in the west. That's kind of my hypothesis here.
   But since it did not get invented there, and in the west they had to invent it and struggle with it and like burning people at the stake and Bruno saying maybe the stars have other worlds it and that's punishable by death. And Galileo, so there's blood on the tracks. So we're inventing science in

a culture and in a philosophy that does not understand what science is trying to do. So it's just an interesting thought experiment, I think!

Dr. Freedman: That's very fascinating. There are a couple; I have a couple of...

Dr. Tyson: And I can't stay much longer. I got a meeting that I'm late for.

Dr. Freedman: Oh I'm sorry.

Dr. Tyson: It doesn't take; they're waiting to come into this office.

Dr. Freedman: Yeah like Einstein. God does not play dice. Was he relatively independent of Judaism?

Dr. Tyson: 100%.

Dr. Freedman: Yeah that's right.

Dr. Tyson: He's viewed Judaism as a superstitious. Judaism the religion is taken from Judaism as...

Dr. Freedman: Culture.

Dr. Tyson: ...culturally, like the sader [Passover dinner]. And people pick the quotes where he used God and they want to believe that he was religious but they're religious and it matters to them that a smart person is religious. But he wasn't religious at all.

Dr. Freedman: What about the Lemaitre?

Dr. Tyson: Lemaitre ... well let me finish with Einstein.

So Einstein, he says God doesn't play dice with the universe. Well it turns out he does. So holding that aside, it's not the first time he invoked God. To him God was just the universe and not an old white man in a white beard who listened to people's prayers and you can have it if he so chooses.

I have other quotes where he is angered by people who takes his other statements as some kind of evidence for his belief in God. And he said this is nonsense. I can show you the quotes.

Dr. Freedman: Yeah, that would be great. Okay...

Dr. Tyson: So Lemaitre – he was approached by powerful members of the Catholic church. I don't think it was the Pope. It might have been Cardinal level people. After he lays out the tenants of the big bang and the expanding universe using the newly freshly minted general relativity.

And they then say ah-ha. Science has proven that God has created the universe. And what that tells me is that religions are really groping for evidence in support of their beliefs. If it ever shows up they just snatch it quickly because they think they understand how tentative their grounds are upon which they stand that they have, that they require that you have to believe it because you're told to believe it and either evidence conflicts with it.

This is very contrary to the urges to discover the natural world. Lemaitre did the only sensible thing when he said no it doesn't say that at all. And he completely, he rejected the notion that somehow he discovered the hand of God as a creator. Plus of course the evidence shows more it was not created in six days. It took somewhat longer.

Dr. Freedman: Yeah, but anyway, it was such an honor to meet you.

Dr. Tyson: Okay, well my hope this is useful to you at all. But there's a lot to think about and write about.

Personally I think the connections are over reached for where they say people try to connect quantum mechanics with ... there are good books on this.

Dr. Freedman: Yeah – The Tao of Physics – I met Fritjof Capra in the 70's.

Dr. Tyson: I think they over reach. I don't... come back with a prediction. Until then...

Dr. Freedman: Oh I just wanted to mention, Dalai Lama is working with Tibetan masters on that finding areas of the brain light up in certain meditative states.

Dr. Tyson: Oh sure. That's a huge but interesting frontier.

Dr. Freedman: Oh I love that. anyway that's a good place to stop. I really appreciate you taking the time to see me.

Dr. Tyson: I'm glad it worked out. Good luck in your writing.

14 MINDLESS CONSUMPTION

If you read the conversation closely, Dr. Tyson found nothing significant that Buddhist thought could contribute to cosmology. His opinions and insights were totally valid and I respect them to the fullest.

I left the interview with the feeling that we had communicated well about the issues we discussed, and left a lot of material on the table. I learned a lot from this conversation and I hope it is of value to you.

One of the most striking takeaways was the discussion on interdependence, or *interbeing* as Zen Master Thich Nhat Hanh calls it. Dr. Tyson's point of view was that until the Industrial Revolution, there was limited interaction between cultures and peoples and they were isolated enough and not so aggressive on the environment as to impact the ecology of the Earth.

The Buddha recognized interdependence as one of the key factors of life. His teachings were as relevant at his time as they are today. Interdependence is one of the factors of existence.

Now, the problems facing the world today have almost reached the stage of being unsolvable. Let's look at some of these problems and recognize the danger that our civilization is in.

One of the most obvious problems is global warming. Scientific evidence for this can be found in climate change, the melting of glaciers all over the world, and the levels of carbon dioxide in the atmosphere. See, for example, *The Sixth Extinction: An Unnatural History* by Elizabeth Kolbert and *Dark Matter and the Dinosaurs: The Astounding Interconnectedness of the Universe* by Lisa Randall.

Climate change is most obvious in recent years due to extensive forest fires, large scale flooding, the strength of tornadoes and hurricanes, loss of species, melting of the polar ice caps, depletion of potable water, rising sea levels and temperatures, and devastating tsunamis. It seems that the news shows one of these types of stories almost every day. Global warming and the changes in the oceanic currents may actually contribute greatly to these catastrophes.

In my article *Is Global Warming Real?*, I reported on the recent work of photographer James Balog and his documentary, *Chasing Ice*.

> "Chasing Ice is the story of one man's mission to change the tide of history by gathering undeniable evidence of our changing planet. Within months of that first trip to Iceland, the photographer conceived the boldest expedition of his life: **The Extreme Ice Survey**. With a band of young adventurers in tow, Balog began deploying revolutionary time-lapse cameras across the brutal Arctic to capture a multi-year record of the world's changing glaciers." [From the preview].

This film brought tears to my eyes. The best thing I know how to do is to tell other people to watch Chasing Ice.

Another very disturbing situation on our planet is the level of carbon dioxide in the atmosphere. Scientists agree that the long-term safe level of carbon dioxide in the atmosphere is 350 parts per million (ppm). Now we have exceeded 400 ppm of carbon dioxide in the atmosphere.

In my article *Three Numbers You Should Know About Global Warming*, I reported that a 2 degree Celsius rise in the average temperature of the Earth is the safe limit. We are approaching this limit now and there is no way to avoid it. These numbers came from 350.org.

350.org also reported that 565 gigatons is the maximum amount of carbon dioxide that we could put in the atmosphere and have a reasonable chance to stay below 2°. A gigaton is a billion tons – 565,000,000,000 tons of CO_2 in the atmosphere! The problem is that we pour 30 gigatons in the air each year, so by 2020, we will exceed this second number.

The third number 350.org reported was 2,795 gigatons is the number of tons of fossil fuels currently on reserve. Pouring this much into the atmosphere is certainly going to lead to the loss of life as we know it.

Our reliance on fossil fuels is the cause of much of the carbon dioxide in the atmosphere. We continue to dig for oil, and we are consuming petroleum products at an extremely rapid rate.

As reported in my article *Zeitgeist Resurrected*, more than 8 barrels of oil are consumed for each barrel of oil found. This is undeniable proof of the unsustainable nature of our current reliance on oil.

Now there is a major effort to extract oil reserves from shale in a process called *fracking*. However, in my article *Five Frightening Facts About Fracking*, I reported that

1. "Fracking" a single well can require more than 1,000,000 gallons of water. This depletes the local groundwater and can dry up nearby creeks. The wastewater produced by fracking contains high levels of radioactivity that wastewater treatment plants are not equipped to treat.
2. Dangerous fracking chemicals are kept secret. In many states, big drilling companies don't have to disclose what chemicals they use in their fracking fluid – the mixture is a "trade secret." Its effects on your health won't be "secret" though, as independent analysts have identified 41 known chemicals in the fracking fluid that are extremely toxic.
3. The "Halliburton Loophole." Thanks to intensive lobbying from then Vice President Dick Cheney, Big Oil & Gas are exempt from the Safe Drinking Water Act and the Clean Air Act.
4. You can light your tap water on fire. When the high explosions used in fracking scatter the rock in which the gas is contained, it can leak into nearby household wells and drinking water. The gas can

not only make you sick and even kill you, but it's also highly explosive.
5. The number of fracking wells is growing at an exponential rate. Fracking is already underway in 28 states. Big Oil & Gas companies are racing to drill more wells before people realize how dangerous this is to their health and safety. No one is safe from the spread of this dangerous drilling practice.

It is extremely urgent that President Obama stop the Keystone XL pipeline. In order to do this, I encourage you to write him a letter and sign the petition.

Another cause of carbon dioxide in the atmosphere is the waste products of animal grazing. So many square miles of rainforests are being cleared to plant crops for animal feed that we are losing the oxygenation of the air by trees. Rainforest soil is not suitable for long-term use for animal grazing used for food. We are destroying the rainforests to create more animal food in such a way that destroys the arability of the land to produce.

Another problem we are facing today is the use of genetically modified organisms (GMO) in food production. These GMO foods are supposed to have benefits in nutritional value and promote greater crop yields. However, there is no evidence that either of these are true. In fact, the only benefits seem to accrue to the owners of the patents of the GMO foods, primarily Monsanto.

In my article _Are You Playing Genetic Roulette?_, I pointed out that the FDA allowed this to happen to us without extensive testing of any kind. GMO foods are a major cause of obesity today and the fact that 70% of all processed foods in the grocery store contain GMOs.

One striking story about GMO crops in the documentary _Genetic Roulette_ was the segment about GMO cotton crops in India. Until GMO cotton was introduced, the cows would scavenge the leftover cotton crops and remained healthy. When the GMO cotton was introduced, the cows were also granted access to the gleaning of the cotton fields. But, in 5 to 7 days, **they all died**!

When I saw this, I was deeply troubled. I have been to India three times and I know how much they value their cows.

The Buddha told a story about a cowherd who lost his cows. He was searching for them when he passed the Buddha and some of his monks and nuns. The cowherd asked, "My cows have gone away and my sesame plants have been eaten by the locusts. I don't know what I am going to do. I may have to kill myself. Did you seen any cows walk by?"

The Buddha felt compassion for this troubled cowherd and responded, "No, we have not seen any cows. You better look in the other direction."

After the cowherd disappeared down the road, he turned to his monks and nuns and said, "Aren't you glad that you have no cows to lose?"

Now, due to GMO foods we are losing cows, pigs, and other domesticated animals by the thousands. We must stop using GMO foods. In order to do this we must know when we are eating them by having them labeled properly. How can we continue to eat something whose long-term effects are completely unknown?

On the other side of the triangle from GMOs and fossil fuels, we find the big drug companies. They have a vested interest in not finding a cure for cancer because surgery, chemotherapy, and radiation therapy are big business. More people make money from sustaining cancer and related illnesses than are killed by them.

When I had cancer sixteen years ago, the gold standard of treatment for bladder cancer was radical cystectomy, that is removal of the bladder and replacement with an artificial one. I found this process so morbid that I opted for a bladder sparing protocol offered by Dr. William U. Shipley of Harvard University and Massachusetts General Hospital.

In any event, the medical establishment got plenty from me in the form of office visits, radiation, and chemotherapy.

I beat them at their own game.

It really makes me angry that cancer cures like Gerson Therapy (see A Cancer Cure – Gerson Therapy) and others (see Suppressed Cancer Cures) are not in more widespread use. The reason: there is no profit for the drug companies. The Gerson method is now in vogue in Japan and interested people can certainly find practitioners who follow Dr. Max Gerson's ideas.

What the drug companies don't want to know and don't want to understand is that they inter-are with the food production system and the fossil fuel distribution world. This Interbeing with the environment establishes our external metabolism.

Zen Master Thich Nhat Hanh tells us that there is no separation between us and the environment. We are in and a part of Mother Earth. The sooner we wake up to the fact, the sooner we will have the motivation to clean up the air, replenish the soil, repair the fresh water system, rebuild the rainforests, and rely on sustainable energy sources.

How can we stop this run-away degradation of our environment?

Mindful consumption may be a good place to start. What is mindful consumption?

A good explanation of mindful consumption comes once again from Zen Master Thich Nhat Hanh in the Fifth Mindfulness Training:

"Aware of the suffering caused by unmindful consumption, I am committed to cultivate good health, both physical and mental, for myself, my family, and my society by practicing mindful eating, drinking, and consuming. I am committed to ingest only items that preserve peace, well-being, and joy in my body, in my consciousness, and in the collective body and consciousness of my family and society. I am determined not to use alcohol or any other intoxicant or to ingest foods or other items that contain toxins, such as certain TV programs, magazines, books, films, and conversations. I am aware that to damage my body or my consciousness with these poisons is to betray my ancestors, my parents, my society, and future generations. I will work to transform violence, fear, anger, and confusion in myself and in society by practicing a diet for myself and for society. I understand that a proper diet is crucial for self-transformation and for the transformation of society."

When I first heard Thich Nhat Hanh speak about mindful consumption I was deeply moved. He said, "If the West stops drinking alcohol by 50%, we could feed the whole world!" I haven't had an alcoholic drink since hearing this fact.

I have made the conscious decision to limit items I purchase and how I spend money. For example, I recently felt a strong need to get a new computer because the one I'm working on was running really slowly. I considered all kinds of Apple computers as well as Windows 8. Then I decided to see what was causing my computer to slow down so much. I discovered that the web browser I was using required a minimum of 10 processes to simply access email. Since I stopped using that browser, my computer is running much better and I can continue to postpone my decision.

Another example relates to GMO foods. By buying fresh organic fruits and vegetables and sustainable fish from the farmer's market, I successfully avoid GMO foods. I don't eat "junk food" and refuse processed food, crackers, and other snacks in favor of fresh nuts, fruits, and chocolate.

Dr. Jerome Freedman

15 DON'T DISPAIR – TAKE ACTION

The contents of the previous chapter raised a lot of questions and only provided limited answers. Some of the problems may be insurmountable under the present conditions. We should not despair. We should take action.

In many articles on my blog, *Meditation Practices for Healing and Well-being*, I have offered suggestions for steps we can take to solve the problems of mindless consumption in the previous chapter. In this chapter, I will comment on some of these suggestions.

With regard to global warming, you should watch *Chasing Ice* and help spread the word about it by sharing the video with your family and friends.

350.org can also use all the help it can get. Please read *Three Numbers You Should Know About Global Warming* and donating to the cause.

Another thing you can do is to begin to divest yourself of fossil fuel investments as soon as feasible or at least set aside your profits for donation to worthy causes like 350.org.

A while ago, there was an opportunity to co-sign the open letter to President Barack Obama to stop the Keystone pipeline from extensive damage to the environment simply for fracking. You can find the letter at 350.org.

Reducing meat consumption, especially the meat raised on GMO crops can make a significant impact on the environment if enough people do so. This will cause major sellers of fast foods to reduce their reliance on chopping down rainforests for grazing land for the animals they only take to slaughter. This will have the effect of sustaining rainforest lands and reducing the outgassing of nitrous oxide and carbon dioxide.

Foods containing GMOs are already starting to have a tremendous impact on the soil and livestock. You probably read the story about the cows gleaning food from GMO cotton crops in India in the previous chapter. This type of thing is also happening to livestock in the United States. Please urge your Senators and Congress people to pass a bill requiring the labeling of GMO foods. This will ensure that the food people get won't harm their intestinal tracks and deplete their immune system. For more information please see *Are You Playing Genetic Roulette?*

Whenever possible, try to buy organically grown fruits and vegetables. These are less likely to contain GMOs and taste much better. In addition, organic meats should be hormone free.

Another alternative is to buy food that is specifically labeled *Non-GMO*. This is the rule, rather than the exception in most countries in Europe. They are backing away completely from GMO foods and we should be doing the same in America!

I highly recommend that you being some form of meditation practice. The benefits of such a practice have been scientifically proven to improve health and well-being. This subject has been covered very well in my book, *Seven Secrets to Stop Interruptions in Meditation: How to Concentrate and Focus on Your Meditation and Deal with Distractions*.

To get you started, why not try this simple, one minute meditation right now:

Find a place of relative solitude where you can sit down in a comfortable position with your back straight. Set the alarm on your phone or kitchen timer one exactly one minute. With your hands in a relaxed but fixed position, close your eyes and follow your breathing, breath by breath. When the alarm sounds, you can stop and notice how you feel.

Everyone can find at least one minute to do this type of meditation.

16 CLOSING REMARKS

I certainly hope you enjoyed reading this book as much as I enjoyed having the conversation with Dr. Neil deGrasse Tyson and writing about it. This has been a wonderful experience for me as I am a trained software engineer, turned author and meditation teacher.

If you have any questions or comments, you can reach me through my website, **Meditation Practices for Healing and Well-being**:

>http://mountainsangha.org

Please register your book by visiting

>http://mountainsangha.org/re-cosmology/

Thanks!

Dr. Jerome Freedman

ABOUT THE AUTHOR

Dr. Jerome Freedman is an author, healthcare advocate, mindfulness meditation teacher, and a cancer survivor since 1997. He is a long-time practitioner in the tradition of Zen Master Thich Nhat Hanh in which he is an ordained member of the Order of Interbeing. His recent article in The Mindfulness Bell titled "Healthy and Free" touched many people. He is also a certified teacher of the Enneagram in the Oral Tradition with Helen Palmer.

Jerome currently teaches **Mindfulness in Healing** at the Pine Street Clinic in San Anselmo, California and writes daily on his blog, **Meditation Practices**. He is a contributing author of *I Am With You: Love Letters to Cancer Patients*, Nancy Novak, PhD, and Barbara K. Richardson.

Dr. Freedman served on Board of Directors of the Marin AIDS Project and the Advisory Council of the Institute for Health and Healing between 2007 and 2010. He is now a major contributor to the Buddhist Climate Action Network and the Plum Village Climate Response as an activist promoting earth protection.

Dr. Freedman holds a Ph. D. in computer science, along

with two master's degrees in physics and a bachelor's degree in chemical engineering. He still consults internationally on software engineering problems and expert witness cases. He successfully interviewed Dr. Neil deGrasse Tyson on cosmology and Buddhist thought in 2011.

He can be reached by for consultations, dharma talks, lectures, and days of mindfulness by email at jerome [at] mountainsangha [dot] org.

Also by Dr. Freedman
Stop Cancer in its Tracks: Your Path to Mindfulness in Healing Yourself
Seven Steps to Stop Interruptions in Meditation: How to Concentrate and Focus on Your Meditation and Deal with Distractions
Cosmology and Buddhist Thought: A Conversation with Dr. Neil deGrasse Tyson – excerpted by Lion's Roar Buddhist Magazine
The Enneagram: Know Your Type! Awaken Your Potential!
Contributing author to I Am With You: Love Letters to Cancer Patients, edited by Nancy Novak, PhD and Barbara K. Richardson

Guided Meditation Recordings
Anger Control Guided Meditation
Achieve Goals Guided Meditation
Sound Sleep Guided Meditation
Stress Relief Guided Meditation
Reduce Symptoms Guided Meditation
Weight Loss Guided Meditation

Order from mountainsangha.org/products.

ACKNOWLEDGMENTS

I wish to express my gratitude to Dr. Neil deGrasse Tyson for the time he spent with me that late spring day in New York in 2011.
It was a very special honor to have him answer my questions and offer scientific insight from the point of view of cosmology.

I also wish to offer thanks to Thich Nhat Hanh for his encouragement when he said to me, "You are doing very well, yourself!"

Thanks also go to my daughter, Jessica Freedman for editing the text and my friend, Kushi Kullar for his reviews, suggestions, and support.

www.ingramcontent.com/pod-product-compliance
Lightning Source LLC
Chambersburg PA
CBHW071609170526
45166CB00003B/1028